ABOU

Every great product star

different. Before becoming the innovative product in your hand today, Circuit Scribe began as grad student research at the University of Illinois. Once an idea was born all it took was some enthusiastic Kickstarter Backers and hard work to create a full-fledged company located in Austin, TX and Somerville, MA.

The Circuit Scribe rollerball pen has conductive silver ink, enabling you to draw functional circuits the same way you would doodle in a sketchbook. The magnetic modules from your kit snap right onto your circuits, bringing them to life!

The Circuit Scribe pen and modules make it easy for students of all ages to start learning about circuits. Once you have learned the basics, Circuit Scribe allows you to create anything you imagine. It provides students, makers, artists, designers, and engineers with unlimited possibilities.

Sincerely,
The Electroninks Team

Analisa Russo (*author*), Michael Bell (*editor*), Mack Fixler (*editor*), Eric Chan (*editor*), Juan Carlos Noguera (*graphics*), S. Brett Walker, Jennifer Lewis, Steven Shewchuk & Nancy Beardsley

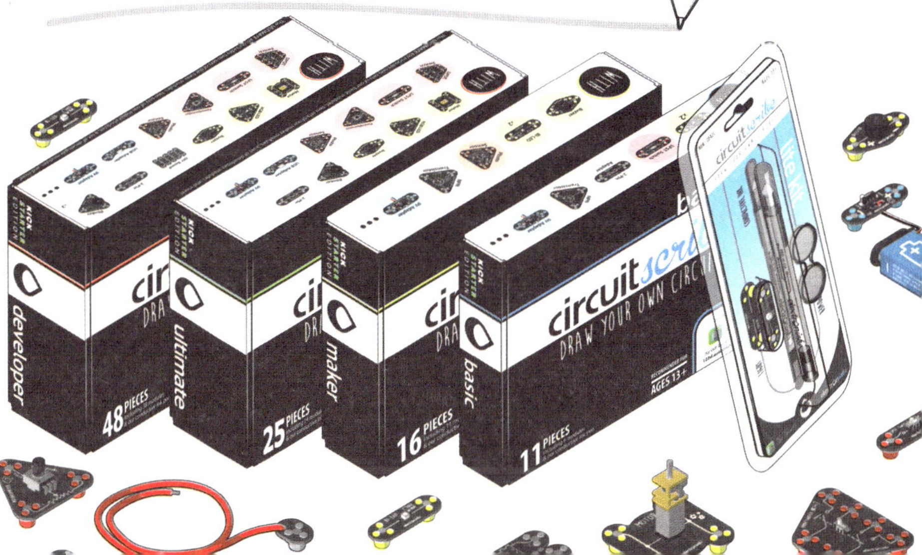

CIRCUIT TIPS

Circuit Scribe is a conductive ink pen. The ink in the pen will allow electricity to flow through the lines that you draw. The drawn lines are ***bare wires.*** Electricity is safe if handled properly. Below are a few tips.

1. ***Circuit Scribe is recommended for ages 8 and up.*** Circuit Scribe kits contain components that could be a choking hazard. Please don't eat the modules or pen ink.

2. It is not recommended to use a power source greater than our 9V battery. Using a larger power source may result in the failure of the ink. Please use Circuit Scribe responsibly.

3. Our 9V battery adapter prevents the battery from short circuiting. If a short circuit occurs, the red light will turn on. Avoid keeping it in this state for a long period of time.

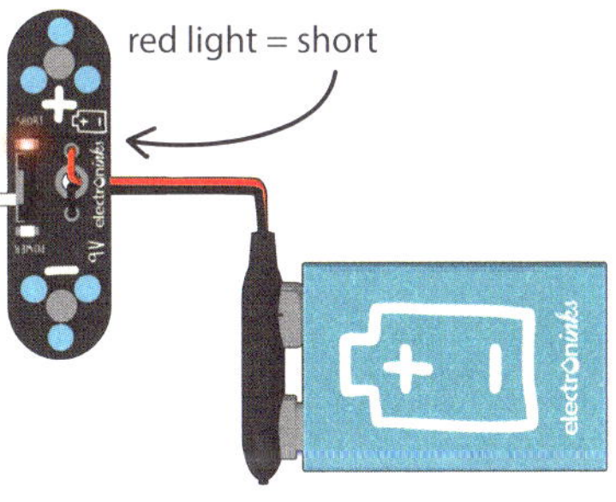

4. ***If the battery appears damaged, or does not output the correct voltage***, discontinue use and dispose of the battery properly. Check local regulations for battery disposal.

5. While Circuit Scribe is certified non-toxic, ***it is not recommended for drawing on skin.*** If you do get silver ink on your skin during normal use, wash it off with soap and water.

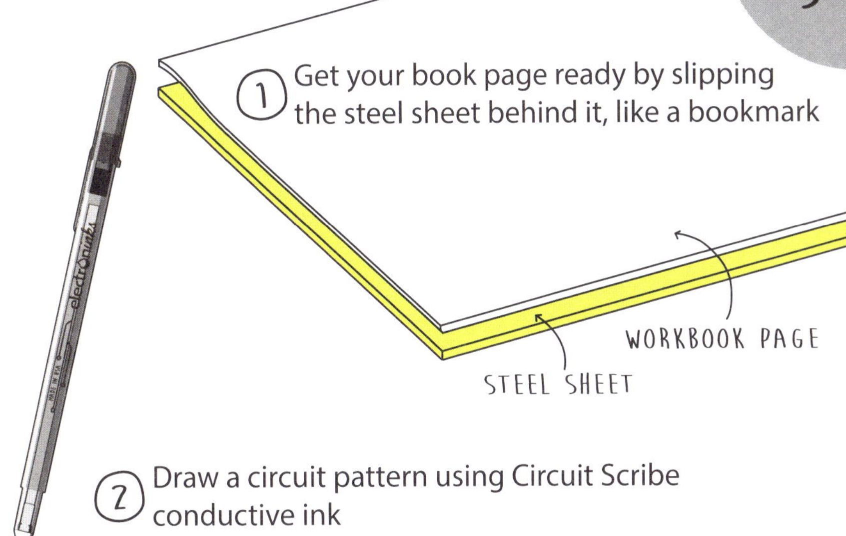

1. Get your book page ready by slipping the steel sheet behind it, like a bookmark

2. Draw a circuit pattern using Circuit Scribe conductive ink

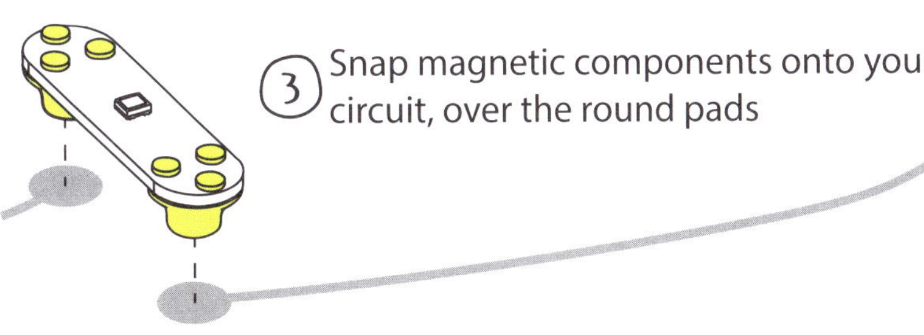

3. Snap magnetic components onto your circuit, over the round pads

HOW THE COMPONENTS WORK

Spherical magnets inside the plastic feet provide a physical connection to the page as well as an electrical connection between the component and the silver ink.

MAGNETIC MODULE COLOR CODE

The colored feet on the modules indicate their function.

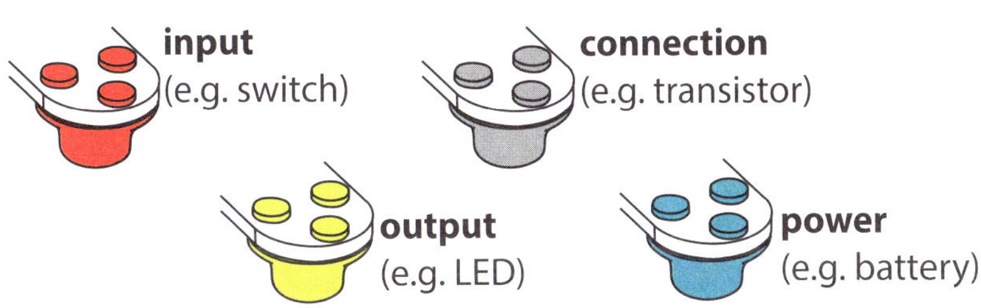

DRAWING IN THE WORKBOOK

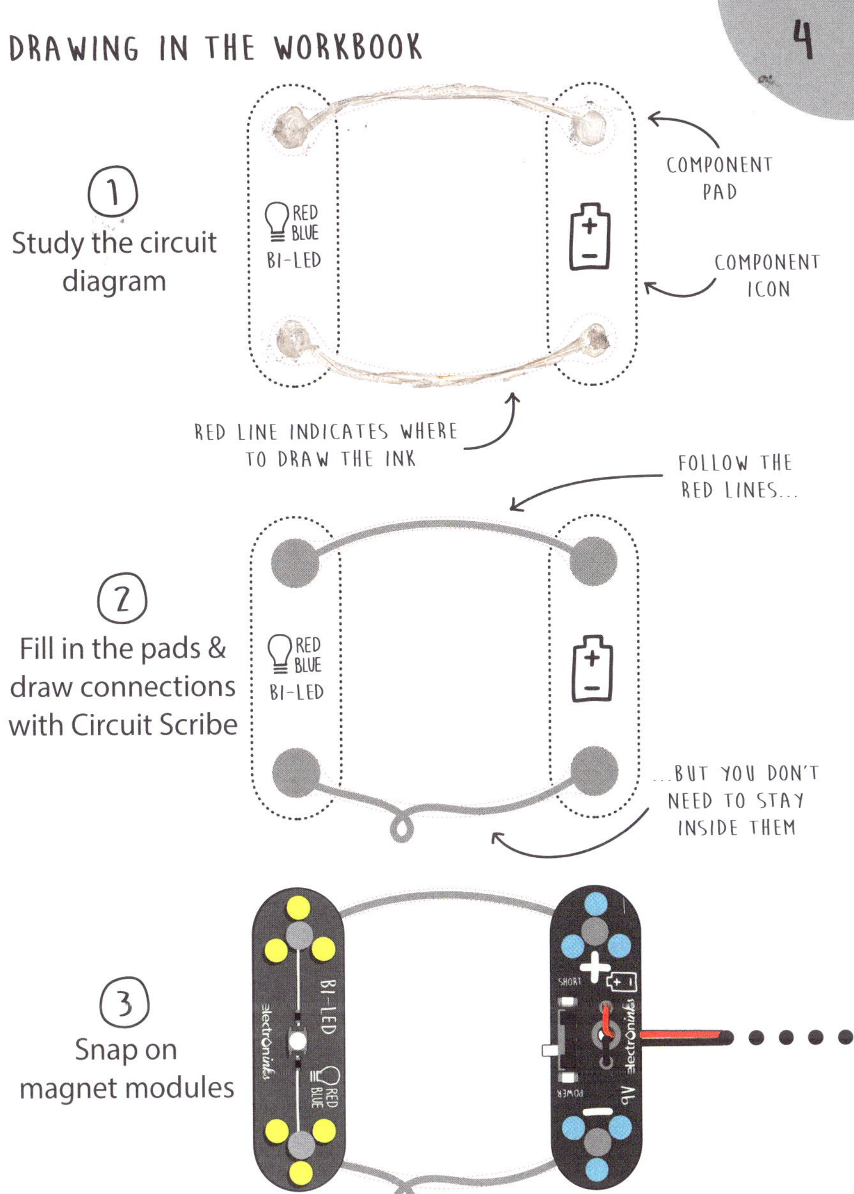

WORKBOOK COLOR CODE

The tabs at the botom of the page show which Circuit Scribe kit you need for the lesson. Sometimes you can substitute an LED for an output module that you don't have. To browse our module library visit ***circuitscribe.com/shop***

B *Basic and above*

M *Maker and above*

U/D *Ultimate or Developer*

CIRCUITS

An electrical circuit is a complete loop that electrons flow through (this flow is called **current)**. Along its path, current can light up **LEDs**, turn **motors**, and sounds **buzzers**. The flow of current can be modified by adding **resistors** or **inputs**.

Now make your first circuit by filling in the pads and connections in the template below.

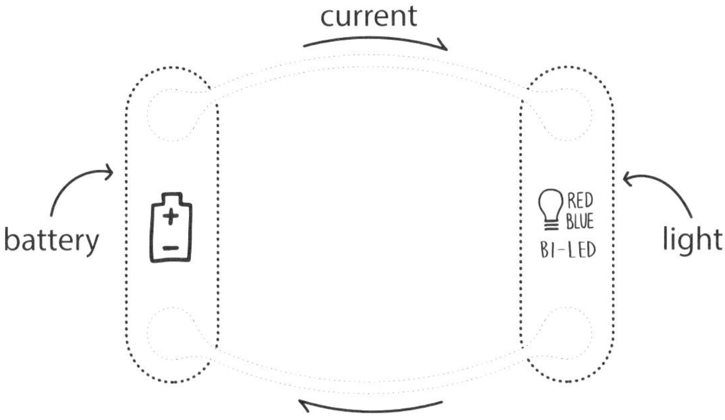

Current is flowing from the postive (+) end of the battery, through the silver ink and LED, to the negative (-) end of the battery.

ELECTRICAL CONDUCTIVITY

Electrically conductive materials allow electrons to flow from one end to the other. Conductivity is a measure of how easily electrons flow through a material. This flow of electrons is called a **current;** its units are called **Amperes** or **Amps** (example: 2.5 A). There are a couple different classifications of conductivity:

Conductor: enables current to flow easily. Circuit Scribe ink is a good conductor.

Insulator: does not allow current to flow at all.

CONDUCTIVITY METER

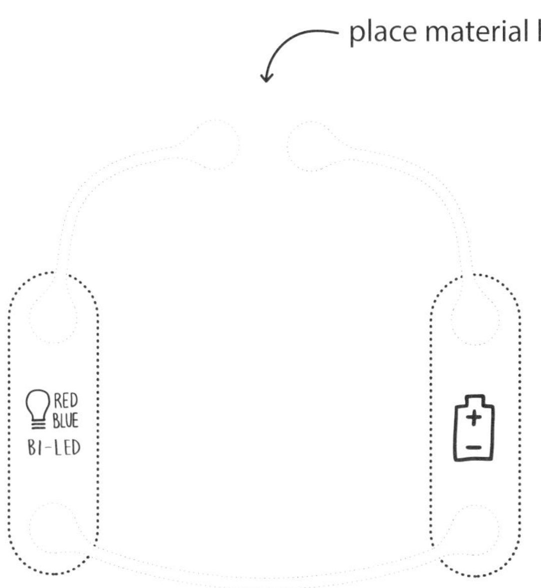

Leave a gap in the circuit above. Press materials from around the house over the pads to complete the circuit.

Using the LED as an indicator, which of these common items are conductors and which are insulators?

paper clip

a pencil

a pipe cleaner

fabric

a bottle cap

rubber band

a key

aluminum foil

refrigerator magnet

mechanical pencil lead

your finger

hair clip

paper

a coin

ELECTRICAL RESISTANCE

A resistor restricts or slows the flow of current through a circuit. Resistors are used to control the brightness of an LED, the volume of a buzzer, or the speed of a motor.

The electrical component called a **resistor** has a resistance value measured in **ohms.** The colored stripes on the resistor are a code that indicates its value. This chart explains how to read the code:

COLOR	1st BAND	2nd BAND	multiplier
BLACK	0	0	1
BROWN	1	1	10
RED	2	2	100
ORANGE	3	3	1,000
YELLOW	4	4	10,000
GREEN	5	5	100,000
BLUE	6	6	1,000,000
VIOLET	7	7	10,000,000
GRAY	8	8	
WHITE	9	9	

Example resistor:

The first two bands are the first two digits (**brown** and **red** = 12).
Multiply by the third band (**blue** = 1,000,000).
The resistor value is 12,000,000 ohms

Resistors in your kit: match the resistor with its value

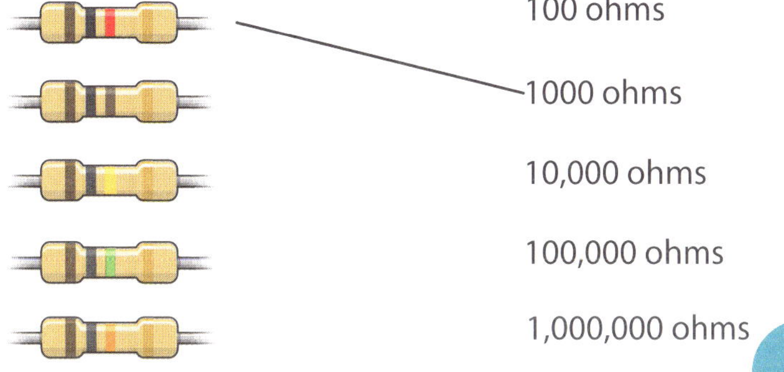

 RESISTORS AND 2-PIN MODULE

This circuit uses the 2-pin module and the 5 resistors that come with your kit. To insert the resistors into the 2-pin module, bend the wires and slide them into the sockets, as pictured below.

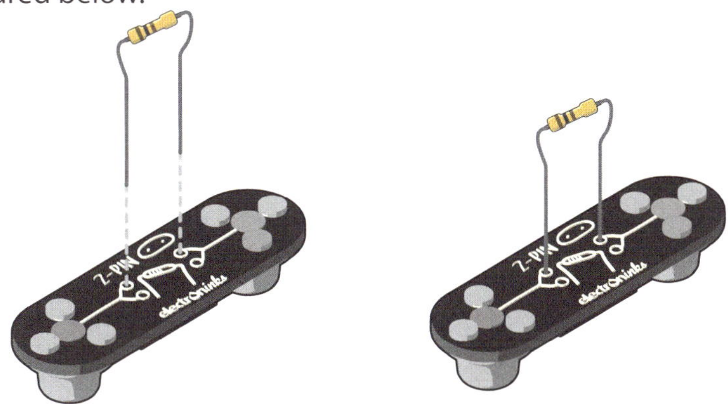

Swap in different resistors after drawing the circuit and observe the change in LED brightness.

Try the photo resistor: This component changes resistance when light is shined on it.

Did you know? *The Circuit Scribe ink has a resistance of about 1-2 ohms per centimeter. The total resistance in the circuit is the resistance of the resistor PLUS the resistance of the ink.*

B

VOLTAGE

Voltage is the change in electric potential energy between two points.

The battery is our energy source. The voltage across the battery terminals is 9 Volts.

In the example circuit below, **current** flows from the positive (+) end of the battery, through the LED, to the negative (-) end of the battery.

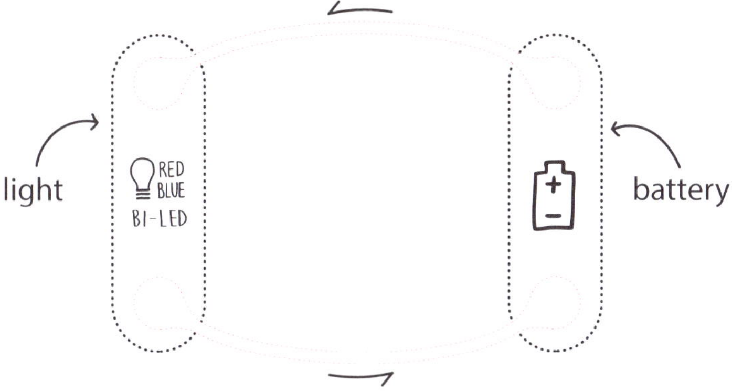

Current takes the path of least resistance. Observe what happens in the circuit below.

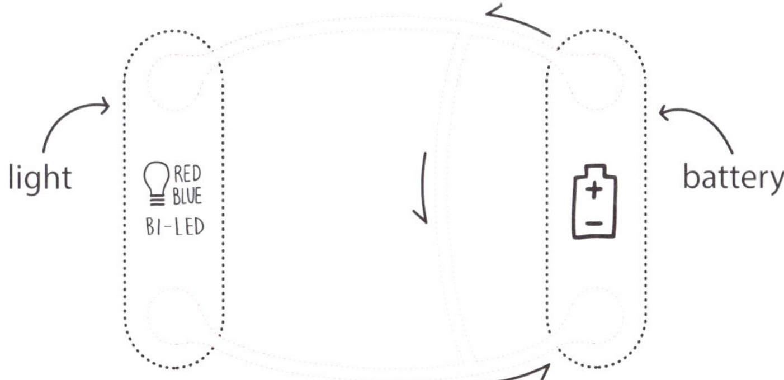

The battery is **"shorted"** and the red indicator light comes on. The two ends of the battery are directly connected by a low-resistance path, so not enough current flows through the LED to turn it on.

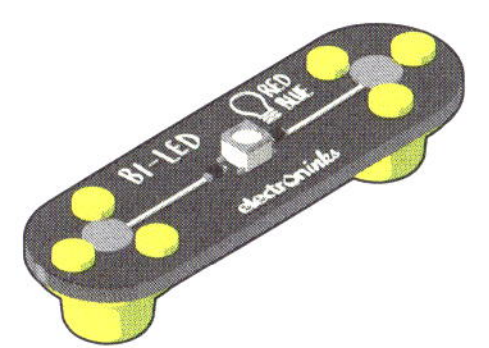

The light emitting diode (LED) lights up when a large enough voltage is applied across its positive and negative electrodes.

The LED is a type of "diode," which is like a one-way street for current flow. Current only flows in the direction of the diode arrow symbol, and is blocked in the opposite direction.

Usually LEDs only work in one direction. Our LED modules have 2 LEDs wired up in opposite directions, which lets you flip the LED around to alternate its color between blue to red.

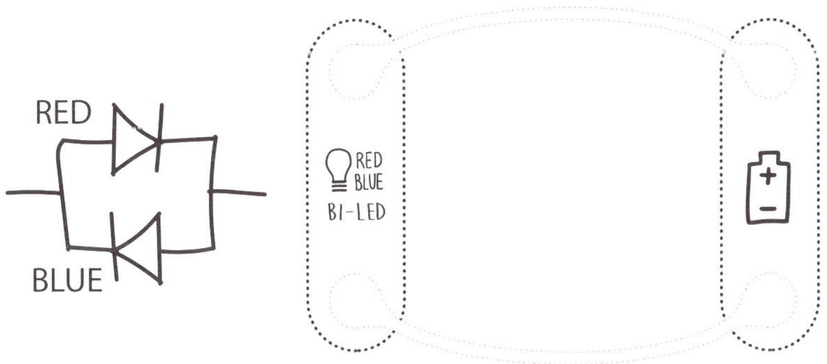

After sketching the circuit, snap on the battery and LED modules. Then flip the direction of the LED and see which orientation activates the **red** LED and which activates the **blue** LED.

SINGLE-POLE SINGLE-THROW SWITCH

The single-pole single-throw (SPST) switch is like a regular light switch: in one position, the circuit is **closed** (connected) and in the other position the circuit is **open** (not connected). The LED will turn on and off when you flip the switch.

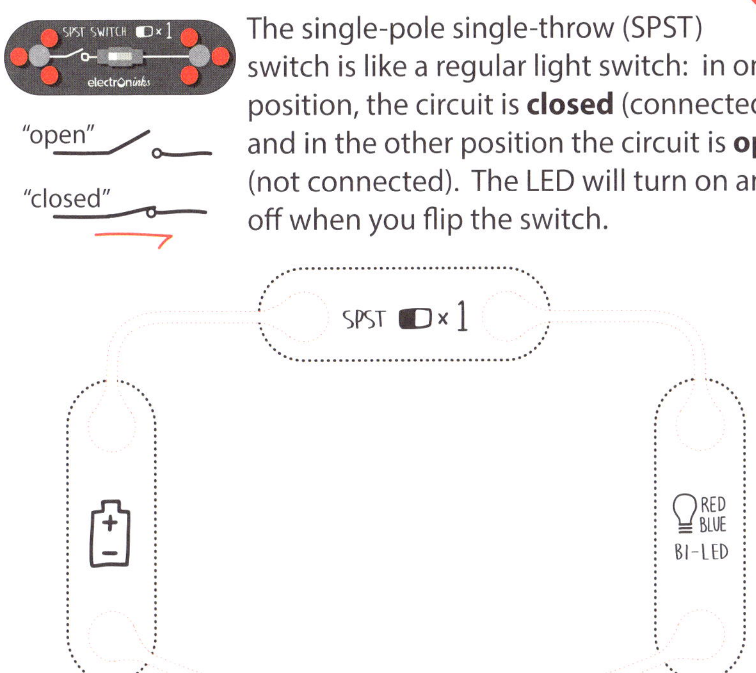

PAPER "PUSH-BUTTON" SWITCH

It's also called a "momentary" switch. The circuit is closed only while you are pressing the button. To make a button using Circuit Scribe and paper, fill in the large oval in the corner and fold the corner over to complete the circuit!

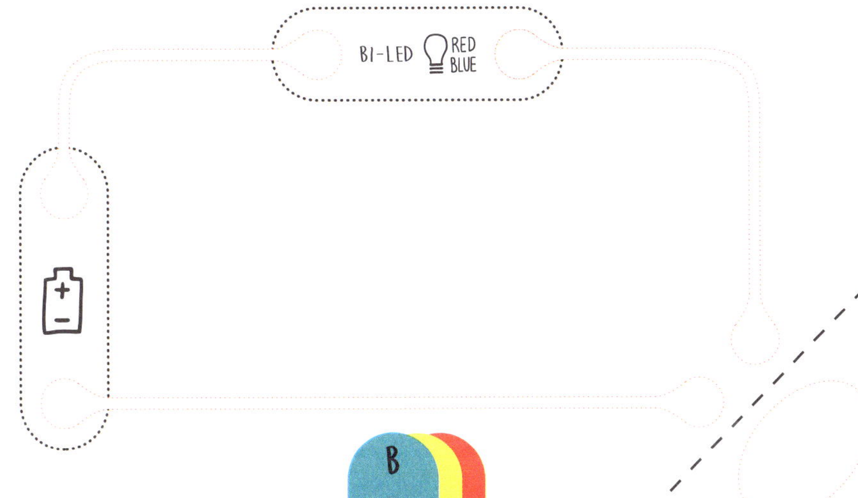

BASIC LOGIC

SWITCHES IN SERIES

In this circuit, both switches (A and B) need to be turned on in order to light the LED.

This is called an **"AND"** gate.

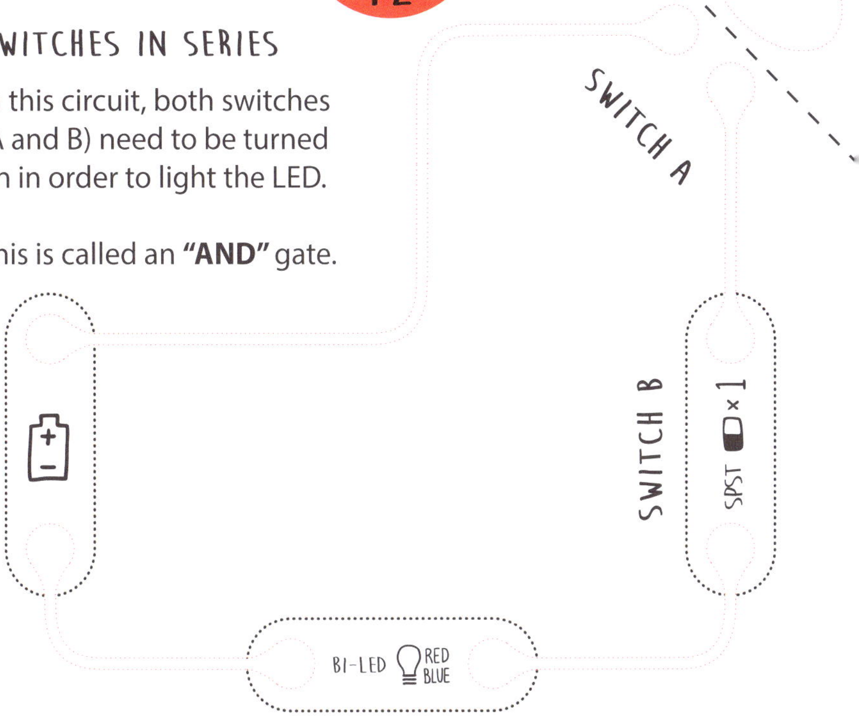

SWITCH A

SWITCH B

SPST × 1

BI-LED RED BLUE

SWITCHES IN PARALLEL

In order to complete the circuit, any combination of switches (A or B or both) can be used to turn on the LED. This is called an **"OR"** gate.

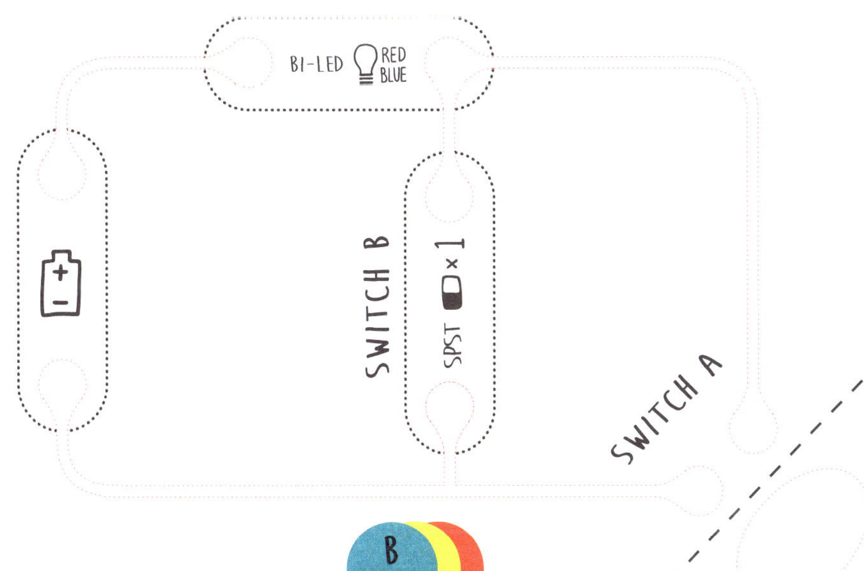

DOUBLE-POLE DOUBLE-THROW SWITCH

The double-pole double-throw (DPDT) switch is used to control and direct the flow of current between the 2 left-hand pads and 4 right-hand pads.

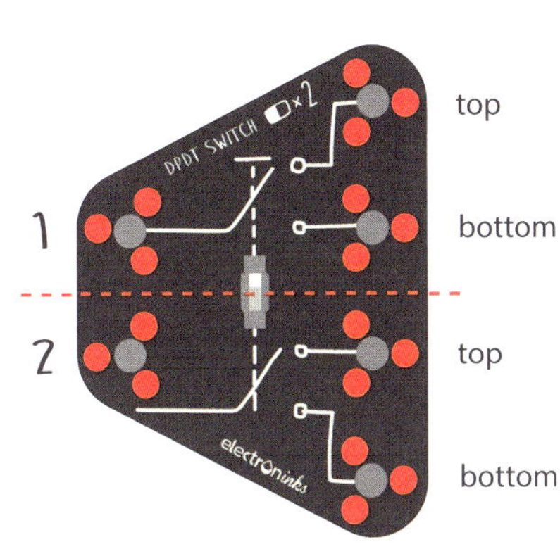

The top half of the switch is called a single-pole double-throw (SPDT).

There is one common pad on the left that is either connected to the top pad or bottom pad on the right.

The DPDT is two SPDT switches operated by one lever!

LED SWAP

Let's use the top half of the DPDT switch as a SPDT swtich. Use it to switch between two LEDs. Can you identify the two separate current paths that exist when you flip the switch?

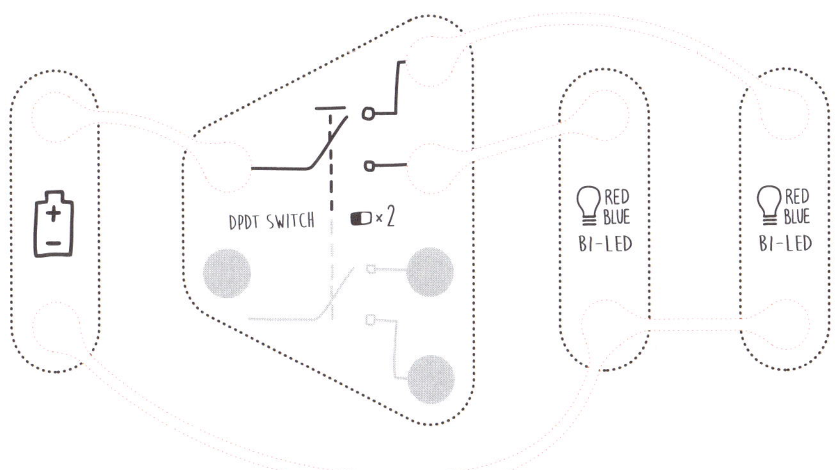

MORE ABOUT THE...
DOUBLE-POLE DOUBLE-THROW SWITCH

When the switch is up, current flows through the two top pads on the right side

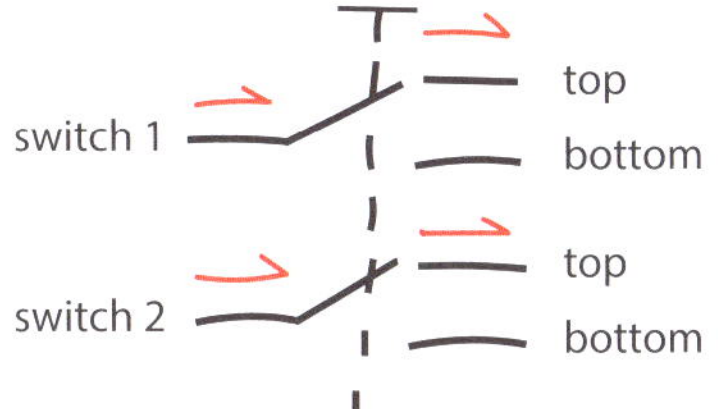

When the switch is down, current flows through the two bottom pads on the right side

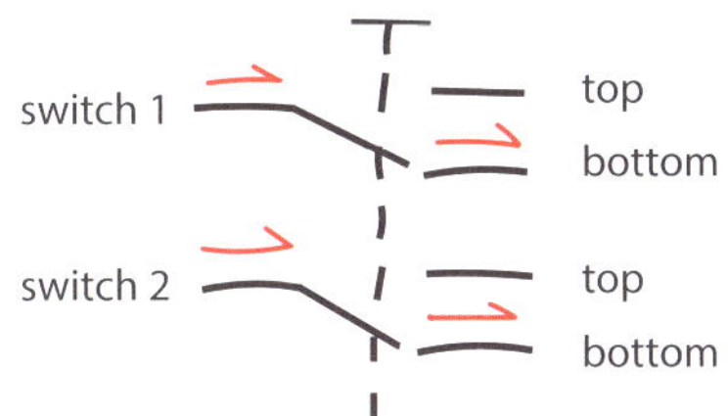

SWITCH LED COLORS

Now use the DPDT switch to change the color on a bi-LED module. The switch is used to direct current flow through the LED in two different directions.

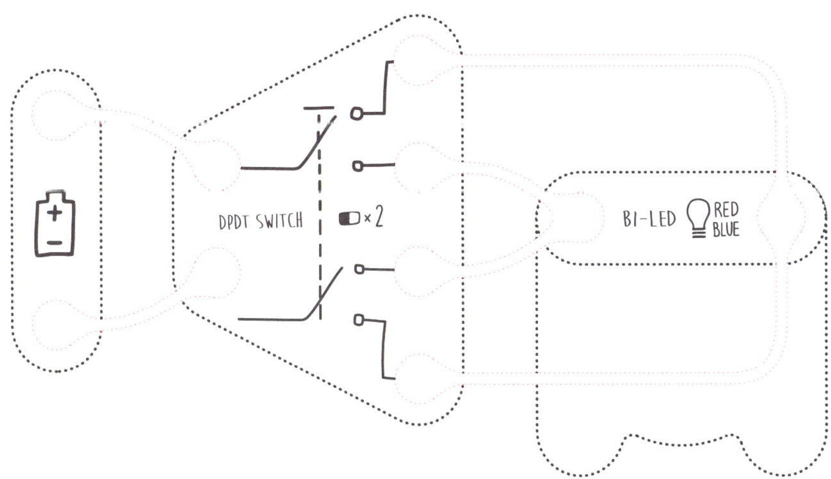

Try the motor! Replace the LED in the circuit above with the motor module. Flip the switch to change the direction of the motor.

NPN TRANSISTOR

The NPN transistor is a current amplifier. Small current between the base and emitter is used to control a larger current between the collector and emitter.

The diagrams below explain how the NPN transistor amplifies a signal.

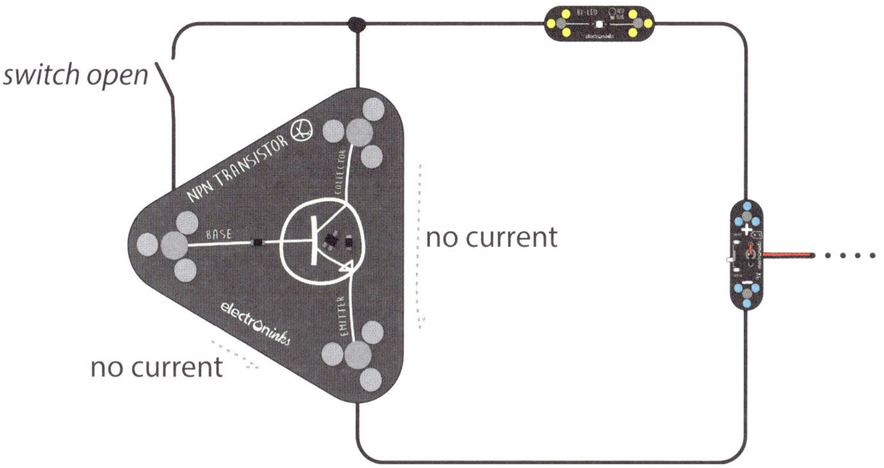

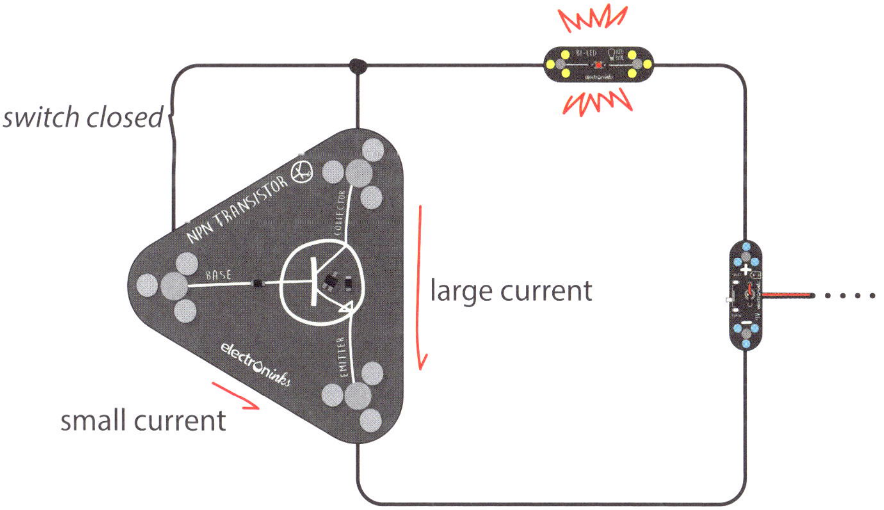

Tip: use the NPN transistor to drive modules that require high current loads, like the motor

Leave a gap between the two pads on the left. After drawing the circuit, touch both pads at the same time. Try using both index fingers.

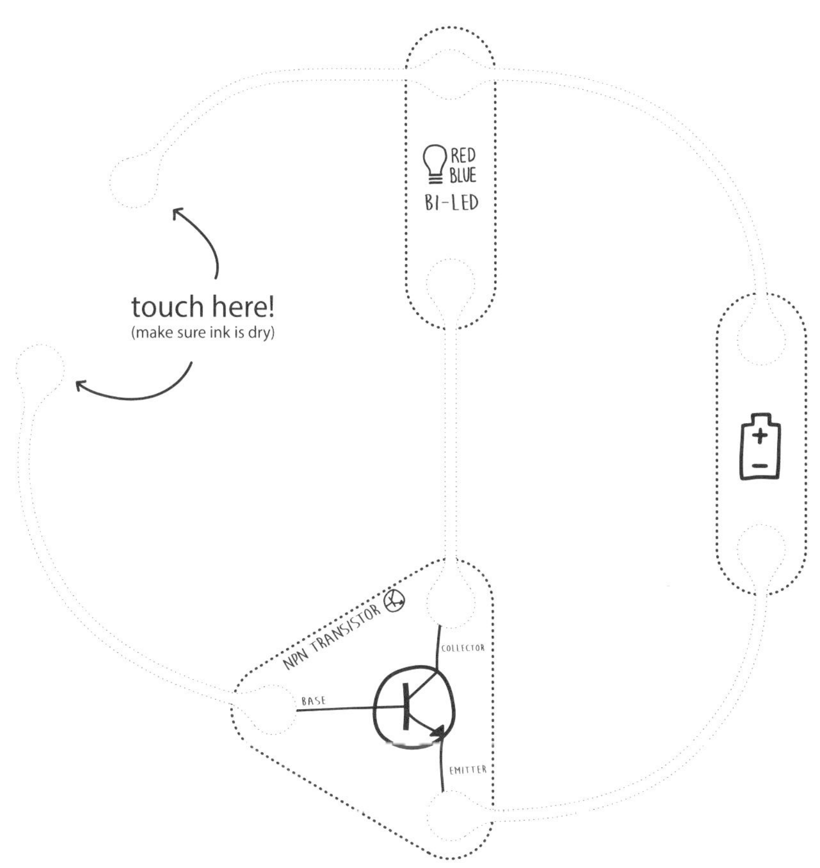

Remember the conductivity meter circuit? Your skin was not conductive enough (or too resistive) to allow the LED to light up. Now it is, because you are using the NPN as an amplifier!

Did you know? *The LED is probably dimly on when you aren't touching the pads. Oil from your fingers and minerals in the paper are conductive enough to activate the NPN transistor and turn on the LED!*

BZZ! BUZZER

The buzzer contains a film that vibrates in response to an electrical voltage. Notice that the buzzer has a PLUS (+) and MINUS (-) sign. The buzzer only works in one direction.

Draw a circuit below using the battery, the buzzer, and the switch.

Now try putting the buzzer and LED **in parallel** with the battery. Once you get it working, remove the LED. Then put the LED back and remove the buzzer. What happens?

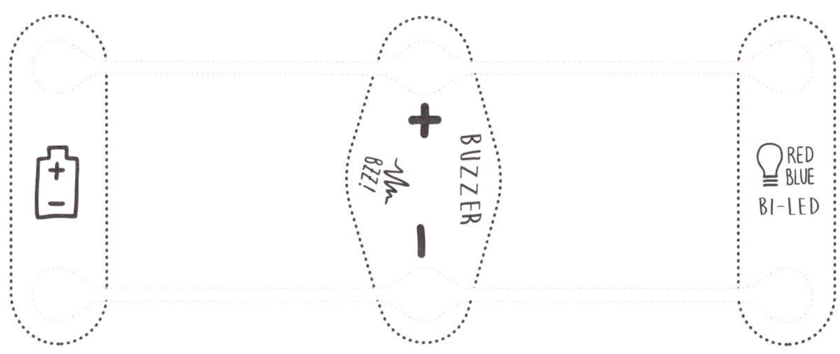

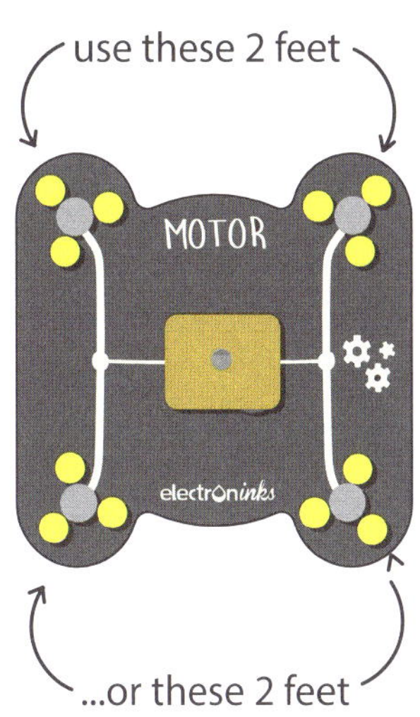

The motor converts electrical energy into rotational motion.

On our module, there are four feet for stability. The two on the left are connected, as are the two on the right.

If your motor is not spinning, try rotating it 90 degrees.

Draw a circuit below using the battery, the motor, and an SPST switch.

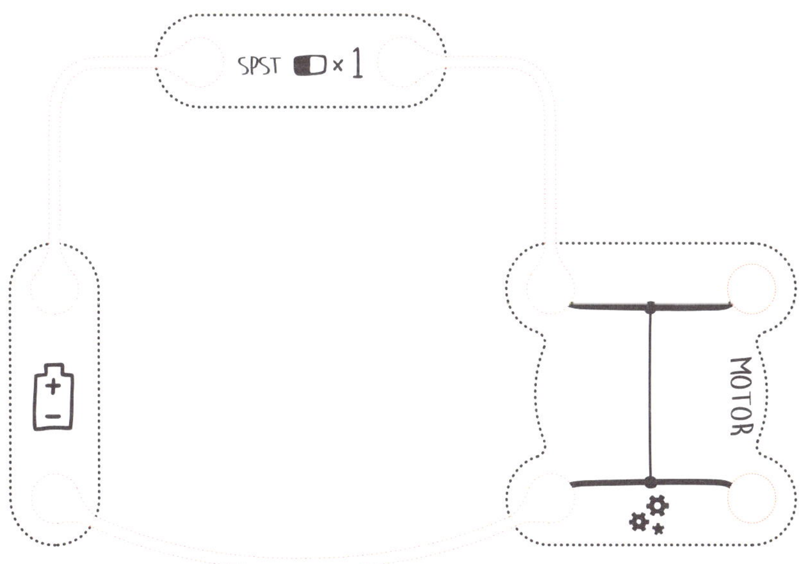

Add a propeller! Attach the propeller from your kit, or make your own propeller or origami to stick on the motor.

RGB LED

The red-green-blue (RGB) LED has 3 colored LEDs packaged on the same module. All three colored LEDs share a negative (ground) electrode, and have separate positive electrodes. You can turn on 1, 2, or 3 different colored LEDs at the same time.

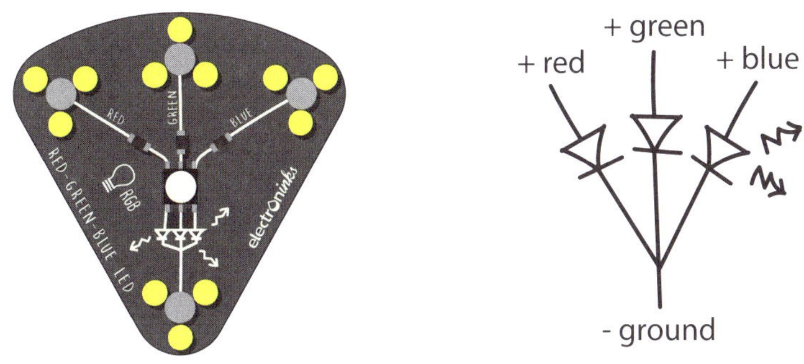

TURQUOISE AND WHITE

In the circuit below, use the switch to mix all three colors together. This produces white light!

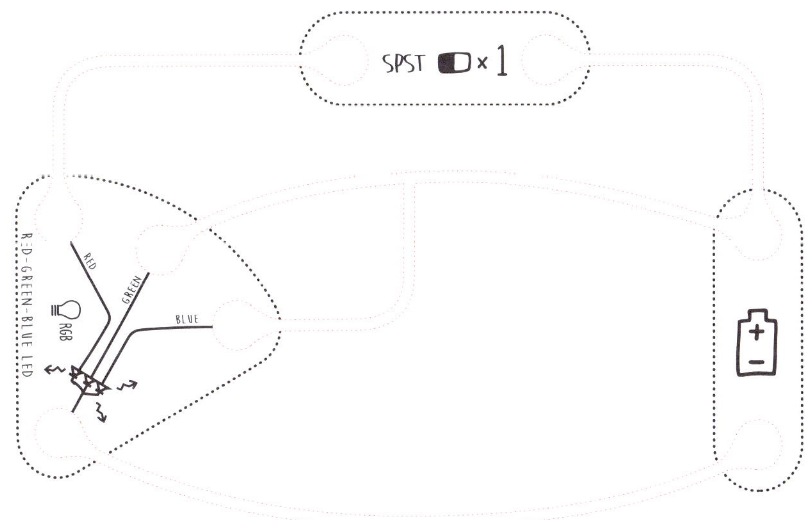

Create a color mixer: on separate paper, try rigging up three switches to operate each color independently.

COLOR PALETTE

Draw connections between open pads where the lines indicate; leave gray pads empty. Keep the ground electrode of the RGB LED fixed on the center pad, then rotate it around the circuit.

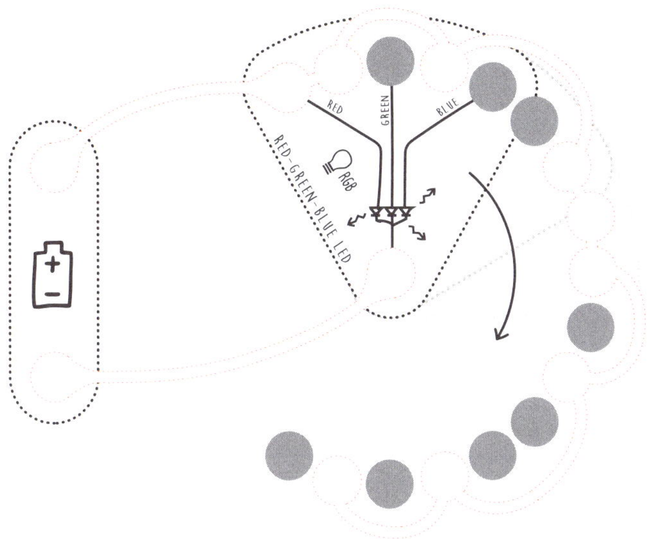

Use the template below to design your own color palette:

If you don't have an RGB LED, make a blinking light! Use a bi-LED module instead, and rotate it around the circuit to make the light blink on and off. Try it with the buzzer, too!

POTENTIOMETER

A potentiometer is a variable resistor. Turning the knob adjusts the amount of resistance between the "wiper" and each of the terminals (feet 1 and 2).

The "wiper" is always used in the circuit: resistance is measured between the wiper and 1 or between the wiper and 2. Let's see what happens when we turn the knob.

All the way clockwise:

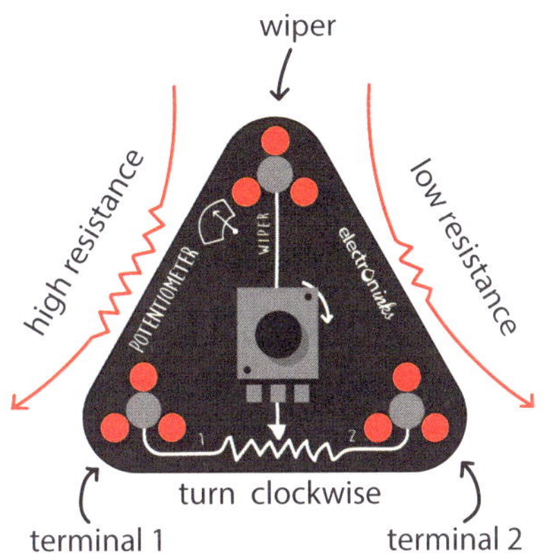

The resistance between wiper and terminal 1 is the highest it can be: 10,000 ohms.

The resistance between the wiper and terminal 2 is very low -- almost zero ohms.

In the middle

Both branches have the same resistance: 5000 ohms. Notice that the sum adds up to 10,000. This is always true regardless of the position of the knob.

Try turning the knob back and forth: you can feel a slight click when it passes through the middle point.

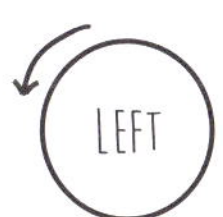

All the way counter-clockwise

This is simply the opposite of the first case (turning clockwise).

The resistance between the wiper and terminal 2 is the highest it can be: 10,000 ohms.

FADER

The circuit below uses the wiper and terminal 2 to control the brightness of an LED. **Variation:** swap the LED for the buzzer and use the potentiometer to change its volume.

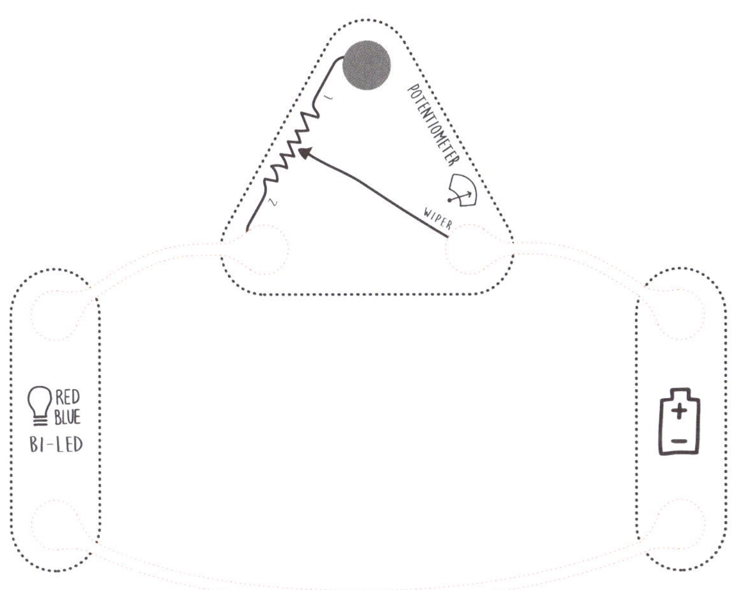

COLOR FADER

Use the potentiometer to transition between the green and blue LEDs on the RGB LED.

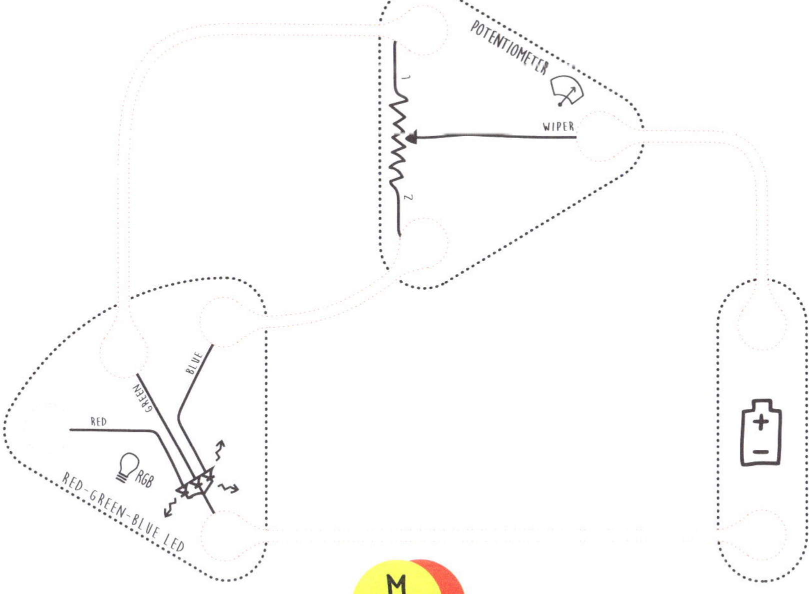

The photo sensor module uses an NPN photo transistor. An NPN photo transistor activates its base with light instead of with electricity.

When light is shining on the photo sensor, current flows between the VCC and output feet.

The photo sensor is an **active** module, meaning that it requires power from the battery in order to work.

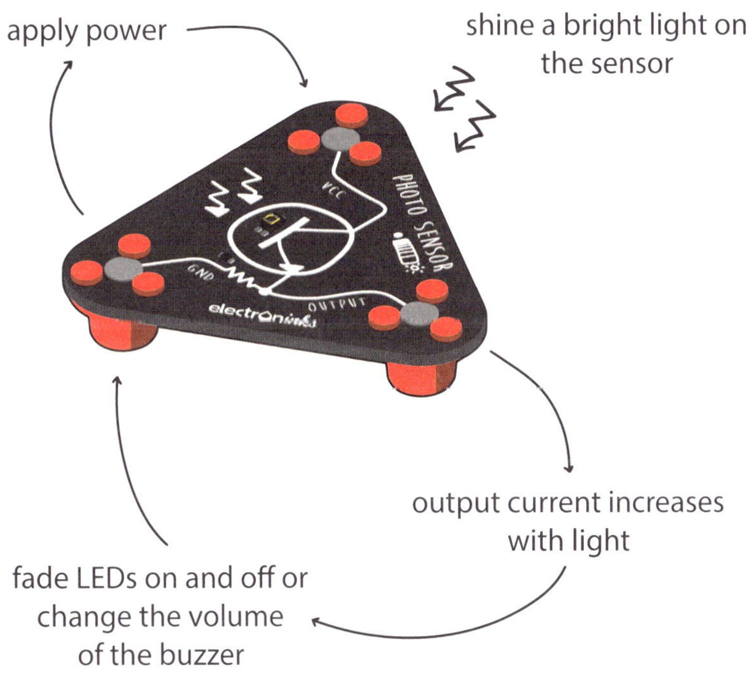

BRIGHT LIGHT SENSOR

Shine a bright light on the photo sensor and see how the brightness of each LED changes. Try it with a **buzzer**, too!

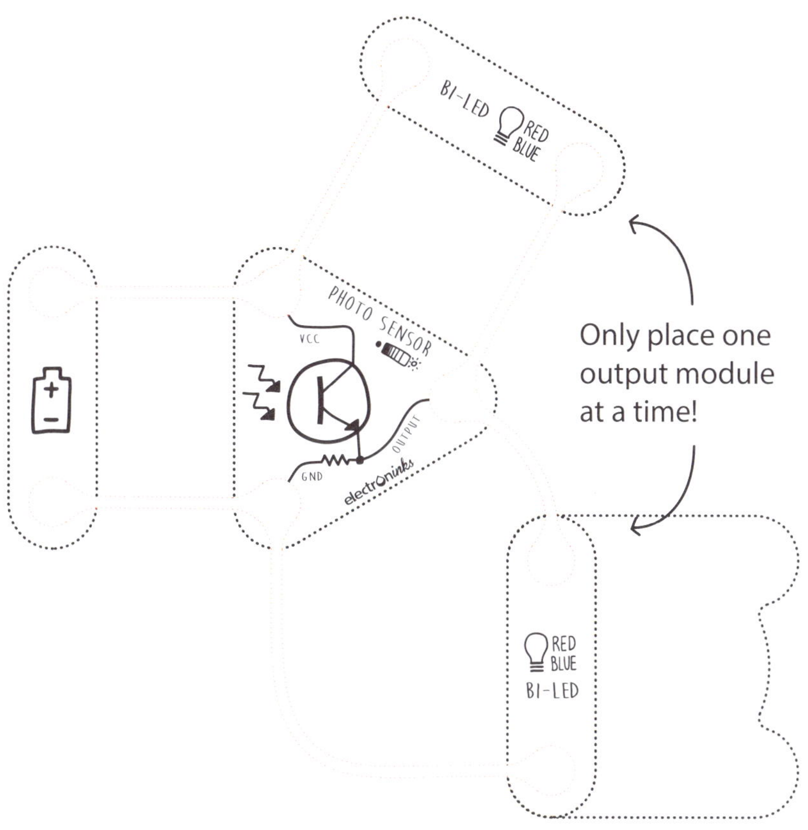

The LEDs in this circuit are connected in two different ways:

1) The bottom LED is connected between the output and ground. When you shine light on the sensor, the LED turns on and gets brighter.

2) The top LED is connected between the positive end of the battery and the output. The LED is normally on, and it gets dimmer when you shine light on the sensor.

Try connecting the motor. It probably won't start turning until your flashlight is very close to the sensor. There is not enough current running through the circuit to turn the motor.

LIGHT-CONTROLLED MOTOR

In the previous example, it was probably hard to get the motor started using the photo sensor. That's because not enough current was running through the output of the photo sensor module.

If we want to control the motor using the photo sensor module, we need to amplify the sensor output. This is a perfect time to use the NPN transistor!

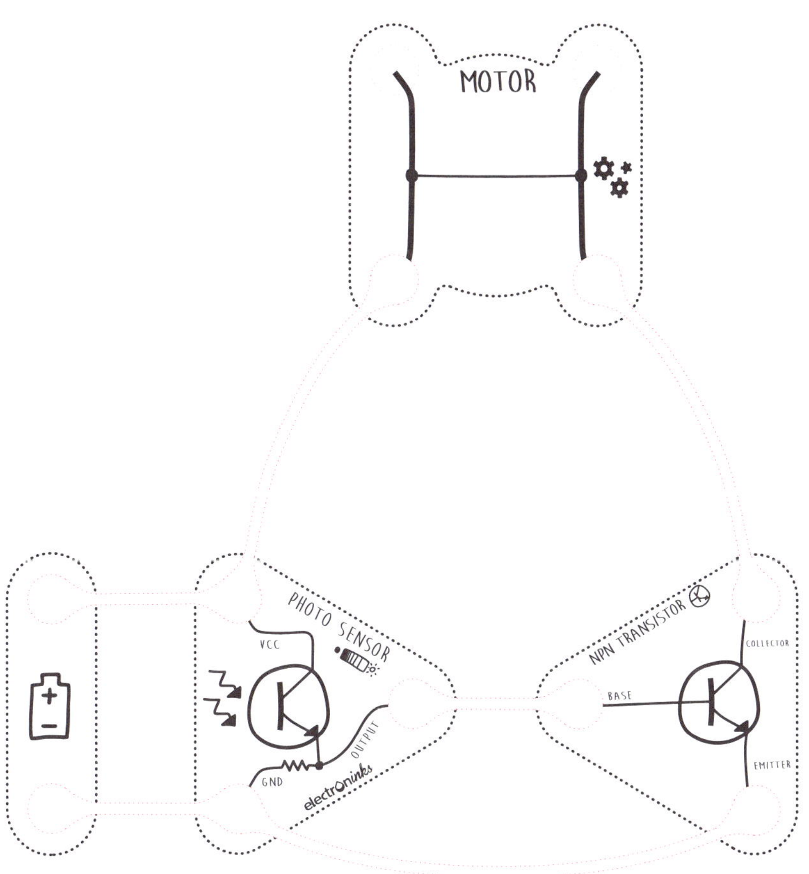

BLINKER

The blinker module uses a chip called a **555 timer.** This module turns its output on and off at a constant rate. Its maximum voltage is the same as your power source (for example, 9 Volts) and its minimum voltage is 0 Volts (ground).

Use this component to blink an LED, beep the buzzer, or turn the motor on and off.

Like the photo sensor, the blinker is an **active** module that requires power from the battery.

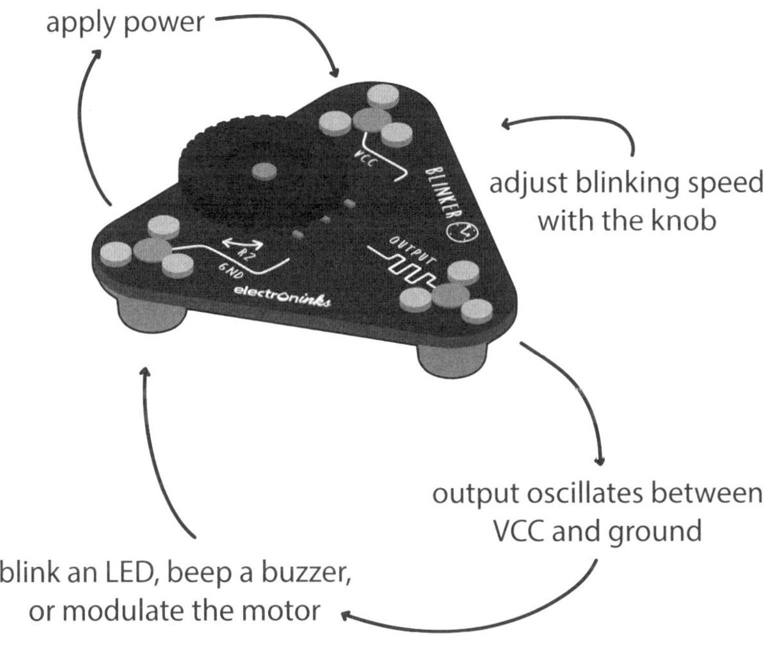

There is a potentiometer on this module. You can change the rate of blinking by turning the dial. There is a green indicator light on the module, as well.

FLASHING LIGHTS

Give the blinker module a try in the circuit below. The two LEDs will flash at the same rate, but when one is on, the other is off. Try turning the knob to change the blinking rate.

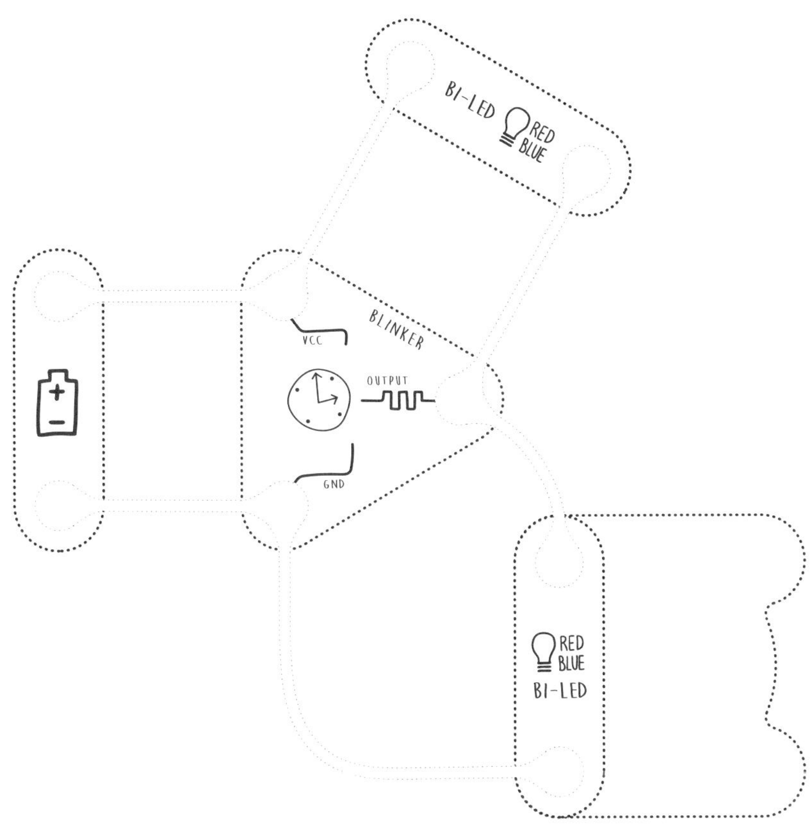

The LEDs in the circuit are connected in two different ways.

1) The bottom LED is connected between the output and ground. The light turns on whenever the output is at 9V. Notice that this LED is in sync with the LED on the Blinker module.

2) The top LED is connected between VCC and the output. The light turns on when the output is low (ground). This LED flashes when the other is turned off.

Try replacing an LED with the buzzer or motor!